THIS BOOK BELONGS TO:

THE WONDERFUL WORLD OF HEDGEHOGS

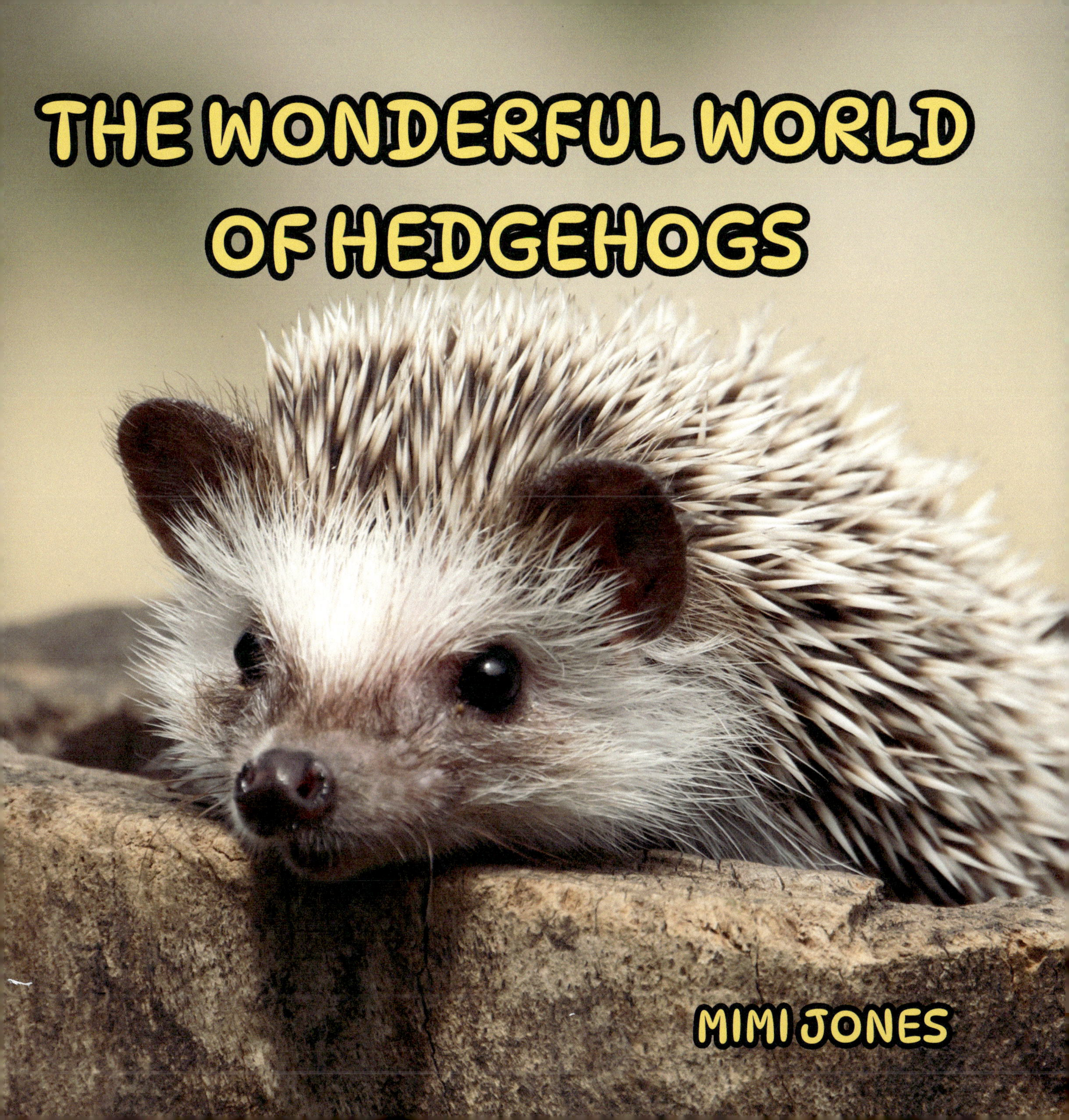

MIMI JONES

Dedicated to all the hedgehog lovers out there.

ISBN 978-1-958985-66-3

www.joeysavestheday.com

A Mimi Book

Hedgehogs are covered in short, stiff spines that are actually hollow hairs made of keratin, which is the same protein found in human hair and nails.

Hedgehogs have 5,000 to 7,000 spines covering their backs! Spines grow and fall out naturally as the hedgehog grows, this shedding is called quilling. Quills are hollow and flexible and usually last up to a year.

They have long noses which help them sniff out food. They can't see very well, but they have a great sense of hearing and smell.

They can run about 4 mph which is pretty quick for a small creature!

Hedgehogs are good swimmers!

SWIM

Hedgehogs are nocturnal, which means they are awake at night and sleep during the day. They can sleep up to 18 hours each day!

All hedgehog species can roll into a ball.

Hedgehogs are mostly solitary, meaning they prefer to be alone rather than with other hedgehogs.

A group of hedgehogs is called an array, though it's very rare to see one.

Hedgehogs are omnivores, they eat bugs, fruits, vegetables, eggs, and even small snakes. They can eat up to one-third of their weight in one night!

Hedgehogs are gardeners' friends, helping them out by eating pests.

Don't give hedgehogs milk! They're lactose intolerant, so water is best.

There are about 17 species of hedgehogs worldwide.

Hedgehogs are native to Europe, Asia, and Africa, and have also been introduced to New Zealand.

Hedgehogs live in grasslands, hedgerows, woods, and meadows.

Mediterranean and grassland hedgehogs can dig burrows as deep as 45 centimeters. Sometimes they will take up residence in abandoned dens left by other animals.

In cold regions, they hibernate in winter to survive freezing temperatures.

Desert hedgehogs sometimes take long summer naps, called aestivation, to stay cool during extreme heat. It is similar to hibernation, but for hot weather instead of cold.

Hedgehogs can live about 2-7 years in the wild, and up to 10 years in captivity.

A baby hedgehog is called a hoglet.

Hoglets are born blind, and their eyes open after a few weeks

Mom hedgehogs can have about 4–7 hoglets in a litter.

Hoglets stay with their mom for about 6-8 weeks before exploring on their own.

In some parts of the U.S., keeping hedgehogs as pets is illegal, such as in California and New York.

African pygmy hedgehogs are the most common type of pet hedgehog. They need special care and are more for looking at than cuddling.

In the UK, hedgehog populations have dropped by about 30% in recent years.

To help them, people make “hedgehog highways” which are small holes in fences so the hedgehogs can roam between gardens.

Count the hedgehogs!

Thank you for exploring the
Wonderful World of Hedgehogs!

Made in the USA
Monee, IL
05 November 2026

67379442R00021